A SIMPLE APPROACH TO LIMIT OF A FUNCTION

A Step-by-Step Review and Practice workbook with Exercises and related solutions on Limit Theorem, Continuity of a Function, Trigonometric Limits, and limit involving Infinity.

I0625446

Adegboye Samuel

TABLE OF CONTENTS

Chapter one

LIMITING VALUE OF A FUNCTION

Understand the Limit of a Function

The two broad areas of calculus known as *differential* and *integral calculus* are built on the foundation concept of a *limit*. In this section our approach to this important concept will be intuitive, concentrating on understanding *what* a limit is using numerical and graphical examples. In the next section, our approach will be analytical, that is, we will use algebraic methods to *compute* the value of a limit of a function.

The limit of a function is a fundamental concept in calculus, which is used to analyze the behavior of a function as it's getting closer to a particular point.

The primary focus of this book is to resolve different problems on Limit of a Function, which will make you get a better understanding of how to solve questions on it quickly.

Let us consider the function

$$F_{(x)} = \frac{2 - x}{3 + x}$$

The value of the functions above $F_{(x)}$ will get closer and closer to $\frac{2}{3}$ as $x \to 0$. that is as x gets closer and closer to zero (0).

Then, it can be summarized

That $F_{(x)} \to \dfrac{2}{3}$ as $x \to 0$.

When you mathematically define this function using limit, we have;

$$\lim_{x \to 0} \frac{2 - X}{3 + X} = \frac{2}{3}$$

Simple and important facts about Limiting Value of a Function

1. If we consider the function

$$f_x = x^2$$

At a value of $x > 2$ that is $(2.0001, 2.001, 2.01, 2.1)$.

$$\text{when } x \to 2, f_x = 2^2 = 4$$

$$\text{when } x \to 2.1, f_x = 2.1^2 = 4.41$$

$$\text{when } x \to 2.01, f_x = 2.01^2 = 4.0401$$

$$\text{when } x \to 2.001, f_x = 2.001^2 = 4.004001$$

$$\text{when } x \to 2.0001, f_x = 2.0001^2 = 4.00040001$$

Observe the trend closely

Observed that $f_x \to 4$ *when* $x \to 2$ from a value of $x > 2$ that is $(2.0001, 2.001, 2.01, 2.1)$.

At a value of $x < 2$ that is **(1.9, 1.99, 1.999, 1.9999)**.

$$f_x = x^2$$

$$\text{when } x \to 2, f_x = 2^2 = 4$$

$$\text{when } x \to 1.9, f_x = 1.9^2 = 3.61$$

$$\text{when } x \to 1.99, f_x = 1.99^2 = 3.9601$$

$$\text{when } x \to 1.999, f_x = 1.999^2 = 3.996001$$

$$when\ x \to 1.9999, f_x = 1.9999^2 = 3.99960001$$

And also, from the function **when x → 2** from the value of $x <$ **2** (1.9, 1.99, 1.999, 1.9999), $f_x \to 4$.

In summary;

$\lim\limits_{x \to 2^+} f_x = x$ is the notation for limiting value f_x as x approaches 2 from the right when $x > 2$. This notation is called a **Right-hand limit of f_x** as x tends to 2.

$\lim\limits_{x \to 2^-} f_x = x$ is the notation for limiting value f_x as x approaches 2 from the left when $x > 2$. This notation is called a **Left-hand limit of f_x** as x tends to 2.

2. The limit can be directly found or can be found by using formulas.

Limit Theorems

I refer to these properties of limit as the laws of limit because they enable us to evaluate the limits of a rather large class of function without resorting to geometric figure or graphs.

Once you are familiar with these properties, the solution to some complex problems on limiting value of a function will be easy.

1. $\lim\limits_{x \to a} k = k$

where **a** and **k** are constant. Limit of a constant is the constant itself.

2. $\lim_{x \to a} k f_x = k \lim_{x \to a} f_x$

    This implies that the limit of the product of a function and a constant is equal to the product of the constant and the limit of the function.

3. $\lim_{x \to a}(f_x \pm g_x) = \lim_{x \to a} f_x \pm \lim_{x \to a} g_x$

    This implies the limit of the sum or difference of two functions is equal to the sum or differences of their respective limits.

4. $\lim_{x \to a}(f_x \cdot g_x) = \left(\lim_{x \to a} f_x \right) \cdot \left(\lim_{x \to a} g_x \right)$

    (.) In mathematics implies **multiplication** or **product**. This implies that the limit of the product of two functions (it can be more than two functions) is equal to the product of their respective limit.

5. $\lim\limits_{x \to a}\left(\frac{f_x}{g_x}\right) = \frac{\left(\lim\limits_{x \to a} f_x\right)}{\left(\lim\limits_{x \to a} g_x\right)}$ provided $\lim\limits_{x \to a} g_x \neq 0$

This implies that the limits of two function are equal to the quotient of their limits provided the limit of the division is not equal to zero.

6. $\lim\limits_{x \to a}(f_x)^n = \left(\lim\limits_{x \to a} f_x\right)^n$

where "**n**" is an integer.

7. $\lim\limits_{x \to a}\left(\sqrt[n]{f_x}\right) = \sqrt[n]{\lim\limits_{x \to a} f_x}$ where "**n**" is positive integer

The laws of limit stated above will help when it comes to solving questions on limits

8. $\lim\limits_{x \to a}\frac{1}{x} = 0$

EXERCISE 1

Write out the properties of Limits without checking the book.

C h a p t e r   T w o

Problems on Limit of a Functions

Miscellaneous Questions and solutions

Example 1

Evaluate

$$\lim_{x \to 0}(7 - 2x + 5x^2 - 4x^3)$$

Solution

$$\lim_{x \to 0}(7 - 2x + 5x^2 - 4x^3)$$

Separate the functions using law 3

$$= \lim_{x \to 0} 7 - \lim_{x \to 0} 2x + \lim_{x \to 0} 5x^2 - \lim_{x \to 0} 4x^3$$

Following law 2, separate the product of a constant and limit of a function

$$= \lim_{x \to 0} 7 - 2\lim_{x \to 0} x + 5\lim_{x \to 0} x^2 - 4\lim_{x \to 0} x^3$$

Replace x with 0, as x → 0

$$= \lim_{x \to 0} 7 - 2\lim_{x \to 0} 0 + 5\lim_{x \to 0} 0 - 4\lim_{x \to 0} 0$$

To solve each limit, note that the limit of a constant is the constant itself,

$$= \lim_{x \to 0} 7 - 2\lim_{x \to 0} 0 + 5\lim_{x \to 0} 0 - 4\lim_{x \to 0} 0$$

$$= 7 - 0 + 0 - 0 = 7$$

$$\lim_{x\to 0}(7 - 2x + 5x^2 - 4x^3) = 7$$

Example 2

Evaluate

$$\frac{\left(\lim_{x\to 0} x^2+5x+9\right)}{\left(\lim_{x\to 0} 2x^2-3x+15 \right)}$$

Solution

$$= \frac{\left(\lim_{x\to 0} x^2 + \lim_{x\to 0} 5x + 9\right)}{\left(\lim_{x\to 0} 2x^2 - 3x + 15 \right)}$$

Separate the functions using law 3

$$= \frac{\lim_{x\to 0} x^2 + \lim_{x\to 0} 5x + \lim_{x\to 0} 9}{\lim_{x\to 0} 2x^2 - \lim_{x\to 0} 3x + \lim_{x\to 0} 15}$$

Following law 2, separate the product of a constant and limit of a function

$$= \frac{\lim_{x\to 0} x^2 + 5\lim_{x\to 0} x + \lim_{x\to 0} 9}{2\lim_{x\to 0} x^2 - 3\lim_{x\to 0} x + \lim_{x\to 0} 15}$$

Replace x with 0, as x → 0

$$= \frac{\lim_{x\to 0} 0^2 + 5\lim_{x\to 0} 0 + \lim_{x\to 0} 9}{2\lim_{x\to 0} 0^2 - 3\lim_{x\to 0} 0 + \lim_{x\to 0} 15}$$

To solve each limit, note that the limit of a constant is the constant itself,

$$= \frac{0+0+9}{0-0+15} = \frac{9}{15} = \frac{3(3)}{3(5)} = \frac{3}{5}$$

Therefore;

$$\frac{\left(\lim_{x\to 0} x^2 + 5x + 9\right)}{\left(\lim_{x\to 0} 2x^2 - 3x + 15\right)} = \frac{3}{5}$$

Example 3

Evaluate

$$\lim_{x\to 6} \left(\frac{x^2 - 36}{x - 6}\right)$$

Solution

$$\lim_{x\to 6} \left(\frac{x^2 - 36}{x - 6}\right)$$

First of all, simplify the equation in the bracket

$$= \lim_{x\to 6} \left(\frac{x^2 - 36}{x - 6}\right)$$

Recall,

$$x^2 - a^2 = (x - a)(x + a)$$

$$= \lim_{x\to 6} \left(\frac{(x^2 - 6^2)}{x - 6}\right)$$

$$= \lim_{x\to 6} \left(\frac{(x - 6)(x + 6)}{(x - 6)}\right)$$

Note,

$$\frac{(x-6)}{(x-6)} = 1$$

$$= \lim_{x \to 6}(x + 6)$$

$$= \lim_{x \to 6} x + \lim_{x \to 6} 6$$

Replace x with 6, as x → 6

$$= \lim_{x \to 6} 6 + \lim_{x \to 6} 6$$

To solve each limit, note that the limit of a constant is the constant itself,

$$= 6 + 6 = 12$$

Therefore;

$$\lim_{x \to 6}\left(\frac{x^2 - 36}{x - 6}\right) = 12$$

Example 4

Evaluate

$$\lim_{x \to a}\left(\frac{4x^3 + 2x^2 + x + 2}{x^3 + 2x + 10}\right)$$

Solution

$$\lim_{x \to a}\left(\frac{4x^3 + 2x^2 + x + 2}{x^3 + 2x + 10}\right)$$

Recall;

$$\lim_{x \to a}\frac{1}{x} = 0$$

$$= \lim_{x \to a}\frac{4x^3 + 2x^2 + x + 2}{x^3 + 2x + 10}$$

Note; $x^3 + 2x + 10 = x^3\left(1 + \frac{2}{x^2} + \frac{10}{x^3}\right)$

$$4x^3 + 2x^2 + x + 2 = x^3\left(4 + \frac{2}{x} + \frac{1}{x^2} + \frac{2}{x^3}\right)$$

$$= \lim_{x \to \alpha} \frac{x^3\left(4 + \frac{2}{x} + \frac{1}{x^2} + \frac{2}{x^3}\right)}{x^3\left(1 + \frac{2}{x^2} + \frac{10}{x^3}\right)}$$

$$= \lim_{x \to \alpha} \frac{\left(4 + \frac{2}{x} + \frac{1}{x^2} + \frac{2}{x^3}\right)}{\left(1 + \frac{2}{x^2} + \frac{10}{x^3}\right)}$$

$$= \frac{\left(\lim_{x\to\alpha}4 + \lim_{x\to\alpha}\frac{2}{x} + \lim_{x\to\alpha}\frac{1}{x^2} + \lim_{x\to\alpha}\frac{2}{x^3}\right)}{\left(\lim_{x\to\alpha}1 + \lim_{x\to\alpha}\frac{2}{x^2} + \lim_{x\to\alpha}\frac{10}{x^3}\right)}$$

$$= \frac{\left(\lim_{x\to\alpha}4 + 2\lim_{x\to\alpha}\frac{1}{x} + \lim_{x\to\alpha}\frac{1}{x^2} + 2\lim_{x\to\alpha}\frac{1}{x^3}\right)}{\left(\lim_{x\to\alpha}1 + 2\lim_{x\to\alpha}\frac{1}{x^2} + 10\lim_{x\to\alpha}\frac{1}{x^3}\right)}$$

Replace x with α , as x → α

$$\frac{\left(\lim_{x\to\alpha}4 + 2\lim_{x\to\alpha}\frac{1}{\alpha} + \lim_{x\to\alpha}\frac{1}{\alpha^2} + 2\lim_{x\to\alpha}\frac{1}{\alpha^3}\right)}{\left(\lim_{x\to\alpha}1 + 2\lim_{x\to\alpha}\frac{1}{\alpha^2} + 10\lim_{x\to\alpha}\frac{1}{\alpha^3}\right)}$$

If $\lim_{x\to\alpha}\frac{1}{x} = 0$, substitute into the equation above.

$$\frac{(4 + 2(0) + 0 + 2(0))}{(1 + 2(0) + 10(0))} = \frac{4}{1} = 4$$

Therefore,

$$\lim_{x \to a} \left(\frac{4x^3 + 2x^2 + x + 2}{x^3 + 2x + 10} \right) = 4$$

Example 5

Evaluate

$$\lim_{x \to 1}(4x^3 + 3x^2 + 2x - 2)$$

Solution

$$\lim_{x \to 1}(4x^3 + 3x^2 + 2x - 2)$$

Separate the functions using law 3

$$= \lim_{x \to 1}4x^3 + \lim_{x \to 1}3x^2 + \lim_{x \to 1}2x - \lim_{x \to 1}2$$

From law 2, separate the product of a constant and limit of a

function

$$= 4\lim_{x \to 1}x^3 + 3\lim_{x \to 1}x^2 + 2\lim_{x \to 1}x - \lim_{x \to 1}2$$

Replace x with 1, as x → 1

$$= 4\lim_{x \to 1}1^3 + 3\lim_{x \to 1}1^2 + 2\lim_{x \to 1}1 - \lim_{x \to 1}2$$

To solve each limit, note that the limit of a constant is the

constant itself,

$$= 4(1) + 3(1) + 2(1) - 2$$

$$= 4 + 3 + 2 - 2 = 7$$

Therefore,

$$\lim_{x \to 1}(4x^3 + 3x^2 + 2x - 2) = 7$$

Example 6

Evaluate

$$\lim_{x \to 0}(x + 3)(3x - 3)(2x + 3)$$

Solution

$$\lim_{x \to 0}(x + 3)(3x - 3)(2x + 3)$$

To evaluate the expression above, we will expand the function first

$$\lim_{x \to 0}(x + 3)(3x - 3)(2x + 3)$$

illustrate each equation with for clarity;

$$(x + 3)(3x - 3)(2x + 3) = (A)(B)(C)$$

Multiply (A) *and* (B) *first*

$$(x + 3)(3x - 3) = x(3x - 3) + 3(3x - 3)$$

$$= (3x^2 - 3x) + (9x - 9)$$

Open the brackets

$$= 3x^2 - 3x + 9x - 9$$

$$= 3x^2 + 6x - 9$$

Therefore;

$$(A)(B) = 3x^2 + 6x - 9$$

Multiply $(3x^2 + 6x - 9)$ *and* (C)

$$(3x^2 + 6x - 9)(2x + 3)$$

$$(3x^2 + 6x - 9)(2x + 3) = 3x^2(2x + 3) + 6x(2x + 3) - 9(2x + 3)$$

$$= 6x^3 + 6x^2 + 12x^2 + 18x - 18x - 27$$

$$= 6x^3 + 18x^2 - 27$$

Having gotten the solution to expression

$$(A)(B)(C)=6x^3 + 18x^2 - 27$$

$$\lim_{x \to 0}(x + 3)(3x - 3)(2x + 3) = \lim_{x \to 0}(6x^3 + 18x^2 - 27)$$

Separate the functions using law 3

$$= \lim_{x \to 0} (6x^3 + 18x^2 - 27)$$

$$= \lim_{x \to 0}6x^3 + \lim_{x \to 0}18x^2 - \lim_{x \to 0}27$$

Following law 2, separate the product of a constant and limit

of a function

$$= 6\lim_{x \to 0}x^3 + 18\lim_{x \to 0}x^2 - \lim_{x \to 0}27$$

To solve each limit, note that the limit of a constant is the

constant itself,

$$6(0) + 18(0) - 27 = 0 + 0 - 27 = -27$$

Therefore,

$$\lim_{x \to 0}(x + 3)(3x - 3)(2x + 3) = -27$$

Example 7

Evaluate

$$\lim_{x \to 3}(x^2 + 9)$$

Solution

$$\lim_{x \to 3}(x^2 + 9)$$

Separate the functions using law 3

$$= \lim_{x \to 3}x^2 + \lim_{x \to 3}9$$

To solve each limit, note that the limit of a constant is the

constant itself,

$$3^2 + 9 = 9 + 9 = 18$$

Therefore;

$$\lim_{x \to 3}(x^2 + 9) = 18$$

Example 8

Evaluate

$$\lim_{x \to 4}\left(\frac{x^2 - 16}{x - 4}\right)$$

Solution

$$\lim_{x \to 4}\left(\frac{x^2 - 16}{x - 4}\right)$$

First of all, simplify the equation in the bracket

$$= \lim_{x \to 4} \left(\frac{x^2 - 16}{x - 4} \right)$$

Recall,

$$x^2 - a^2 = (x - a)(x + a)$$

$$= \lim_{x \to 4} \left(\frac{(x^2 - 4^2)}{x - 4} \right)$$

$$= \lim_{x \to 4} \left(\frac{(x - 4)(x + 4)}{(x - 4)} \right)$$

Note;
$$\frac{(x-4)}{(x-4)} = 1$$

$$= \lim_{x \to 4}(x + 4)$$

$$= \lim_{x \to 4} x + \lim_{x \to 4} 4$$

Replace x with 4, as x → 4

$$= \lim_{x \to 4} 4 + \lim_{x \to 4} 4$$

To solve each limit, note that the limit of a constant is the

constant itself,

$$= 4 + 4 = 8$$

Therefore;

$$\lim_{x \to 4} \left(\frac{x^2 - 16}{x - 4} \right) = 8$$

Example 9

Evaluate

$$\lim_{x\to\frac{1}{3}} \left(\frac{9x^2-1}{3x-1}\right)$$

Solution

$$\lim_{x\to\frac{1}{3}} \left(\frac{9x^2-1}{3x-1}\right)$$

First of all, simplify the equation in the bracket

$$= \lim_{x\to\frac{1}{3}} \left(\frac{9x^2-1}{3x-1}\right)$$

Recall, difference of two square

$$x^2 - a^2 = (x-a)(x+a)$$

$$= \lim_{x\to\frac{1}{3}} \left(\frac{(3^2x^2-1^2)}{3x-1}\right)$$

$$= \lim_{x\to\frac{1}{3}} \left(\frac{(3x-1)(3x+1)}{(3x-1)}\right)$$

Note,

$$\frac{(3x-1)}{(3x-1)} = 1$$

$$= \lim_{x\to\frac{1}{3}}(3x+1)$$

$$= \lim_{x\to\frac{1}{3}}3x + \lim_{x\to\frac{1}{3}}1$$

Replace x with $\frac{1}{3}$, as x $\to \frac{1}{3}$

$$= \lim_{x \to \frac{1}{3}} 3x + \lim_{x \to \frac{1}{3}} 1$$

To solve each limit, note that the limit of a constant is the

constant itself,

$$= 3\left(\frac{1}{3}\right) + 1 = 2$$

Therefore;

$$\lim_{x \to \frac{1}{3}} \left(\frac{9x^2 - 1}{3x - 1}\right) = 2$$

Example 10

Evaluate

$$\lim_{x \to a} \left(\frac{x^2 + 3x + 2}{x^2 - 2x + 7}\right)$$

Solution

$$\lim_{x \to a} \left(\frac{x^2 + 3x + 2}{x^2 - 2x + 7}\right)$$

Recall,

$$\lim_{x \to a} \frac{1}{x} = 0$$

$$= \lim_{x \to a} \frac{x^2 + 3x + 2}{x^2 - 2x + 7}$$

Note; $x^2 + 3x + 2 = x^2\left(1 + \frac{3}{x} + \frac{2}{x^2}\right)$

$$x^2 - 2x + 7 = x^2\left(1 - \frac{2}{x} + \frac{7}{x^2}\right)$$

$$= \lim_{x \to \alpha} \frac{x^2\left(1 + \frac{3}{x} + \frac{2}{x^2}\right)}{x^2\left(1 - \frac{2}{x} + \frac{7}{x^2}\right)}$$

$$= \lim_{x \to \alpha} \frac{\left(1 + \frac{3}{x} + \frac{2}{x^2}\right)}{\left(1 - \frac{2}{x} + \frac{7}{x^2}\right)}$$

$$= \frac{\left(\lim_{x \to \alpha} 1 + \lim_{x \to \alpha} \frac{3}{x} + \lim_{x \to \alpha} \frac{2}{x^2}\right)}{\left(\lim_{x \to \alpha} 1 - \lim_{x \to \alpha} \frac{2}{x} + \lim_{x \to \alpha} \frac{7}{x^2}\right)}$$

$$= \frac{\left(\lim_{x \to \alpha} 1 + 3\lim_{x \to \alpha} \frac{1}{x} + 2\lim_{x \to \alpha} \frac{1}{x^2}\right)}{\left(\lim_{x \to \alpha} 1 - 2\lim_{x \to \alpha} \frac{1}{x} + 7\lim_{x \to \alpha} \frac{1}{x^2}\right)}$$

Replace x with α , as x → α

$$= \frac{\left(\lim_{x \to \alpha} 1 + 3\lim_{x \to \alpha} \frac{1}{x} + 2\lim_{x \to \alpha} \frac{1}{x^2}\right)}{\left(\lim_{x \to \alpha} 1 - 2\lim_{x \to \alpha} \frac{1}{x} + 7\lim_{x \to \alpha} \frac{1}{x^2}\right)}$$

If $\lim_{x \to \alpha} \frac{1}{x} = 0$, substitute into the equation above.

$$\frac{(1 + 3(0) + 2(0))}{(1 - 2(0) + 7(0))} = \frac{1}{1} = 1$$

Therefore,

$$\lim_{x \to \alpha}\left(\frac{x^2 + 3x + 2}{x^2 - 2x + 7}\right) = 1$$

Example 11

Find the limiting value of $g_x = (x - 2)(x + 2)$ as x approaches 3.

Solution

$$g_x = (x - 2)(x + 2)$$

$$= \lim_{x \to 3}(x - 2)(x + 2)$$

Expand the expression;

$$= \lim_{x \to 3}(x - 2)(x + 2) = x(x + 2) - 2(x + 2)$$

$$= \lim_{x \to 3} x^2 + 2x - 2x - 4$$

$$= \lim_{x \to 3}(x^2 - 4)$$

$$= \lim_{x \to 3} x^2 - \lim_{x \to 3} 4$$

Replace x with 3, as x → 3

$$= 3^2 - 4 = 9 - 4 = 5$$

Example 12

Evaluate

$$\lim_{x \to a}\left(\frac{5x+1}{x}\right)$$

Solution

$$\lim_{x \to a}\left(\frac{5x + 1}{x}\right)$$

Simplify the function

$$\lim_{x \to a} \left(\frac{5x + 1}{x} \right) = \lim_{x \to a} \left(\frac{5x}{x} + \frac{1}{x} \right)$$

Separate the functions using law 3

$$\lim_{x \to a} \left(\frac{5x}{x} + \frac{1}{x} \right) = \lim_{x \to a} \left(5 + \frac{1}{x} \right)$$

$$= \lim_{x \to a} 5 + \lim_{x \to a} \frac{1}{x}$$

Recall,

$$\lim_{x \to a} \frac{1}{x} = 0$$

To solve each limit, note that the limit of a constant is the constant itself,

$$= 5 + 0 = 5$$

$$\lim_{x \to a} \left(\frac{5x + 1}{x} \right) = 5$$

Example 13

Evaluate

$$\lim_{x \to a} \left(\frac{3x^3 + 2x^2 + x + 1}{x^3 + 2x + 5} \right)$$

Solution

$$\lim_{x \to a} \left(\frac{3x^3 + 2x^2 + x + 1}{x^3 + 2x + 5} \right)$$

Recall,

$$\lim_{x \to a} \frac{1}{x} = 0$$

$$= \lim_{x \to a} \frac{3x^3 + 2x^2 + x + 1}{x^3 + 2x + 5}$$

Note; $x^3 + 2x + 5 = x^3\left(1 + \frac{2}{x^2} + \frac{5}{x^3}\right)$

$$3x^3 + 2x^2 + x + 1 = x^3\left(3 + \frac{2}{x} + \frac{1}{x^2} + \frac{1}{x^3}\right)$$

$$= \lim_{x \to a} \frac{x^3\left(3 + \frac{2}{x} + \frac{1}{x^2} + \frac{1}{x^3}\right)}{x^3\left(1 + \frac{2}{x^2} + \frac{5}{x^3}\right)}$$

$$= \lim_{x \to a} \frac{\left(3 + \frac{2}{x} + \frac{1}{x^2} + \frac{1}{x^3}\right)}{\left(1 + \frac{2}{x^2} + \frac{5}{x^3}\right)}$$

$$= \frac{\left(\lim_{x \to a} 3 + \lim_{x \to a} \frac{2}{x} + \lim_{x \to a} \frac{1}{x^2} + \lim_{x \to a} \frac{1}{x^3}\right)}{\left(\lim_{x \to a} 1 + \lim_{x \to a} \frac{2}{x^2} + \lim_{x \to a} \frac{5}{x^3}\right)}$$

$$= \frac{\left(\lim_{x \to a} 3 + 2\lim_{x \to a} \frac{1}{x} + \lim_{x \to a} \frac{1}{x^2} + \lim_{x \to a} \frac{1}{x^3}\right)}{\left(\lim_{x \to a} 1 + 2\lim_{x \to a} \frac{1}{x^2} + 5\lim_{x \to a} \frac{1}{x^3}\right)}$$

Replace x with α, as $x \to \alpha$

$$\frac{\left(\lim_{x \to a} 3 + 2\lim_{x \to a} \frac{1}{\alpha} + \lim_{x \to a} \frac{1}{\alpha^2} + \lim_{x \to a} \frac{1}{\alpha^3}\right)}{\left(\lim_{x \to a} 1 + 2\lim_{x \to a} \frac{1}{\alpha^2} + 5\lim_{x \to a} \frac{1}{\alpha^3}\right)}$$

If $\lim_{x \to a} \frac{1}{x} = 0$, substitute into the equation above.

$$\frac{(3 + 2(0) + 0 + 0)}{(1 + 2(0) + 10(0))} = \frac{3}{1} = 3$$

Therefore,

$$\lim_{x \to a} \left(\frac{3x^3 + 2x^2 + x + 1}{x^3 + 2x + 5} \right) = 3$$

Example 14

Evaluate

$$\lim_{x \to 4} 5\sqrt{x^2 + 9}$$

<u>Solution</u>

$$\lim_{x \to 4} 5\sqrt{x^2 + 9}$$

Following law 2, separate the product of a constant and limit of a function

$$5 \lim_{x \to 4} \sqrt{x^2 + 9}$$

Replace x with 0, as x → 4

$$5\left(\sqrt{4^2 + 9}\right) = 5(\sqrt{25}) = 5 \times 5 = 25$$

Therefore;

$$\lim_{x \to 4} 5\sqrt{x^2 + 9} = 25$$

Exercise 2: Try this on your own

Evaluate:

1. $\lim\limits_{x \to 0}(6x^5 - 5x^4 + 3x^3 + 5x^2 - x - 3 + a)$

2. Evaluate $\lim_{x \to 0}(x^3 - 2x^2 + 4)$

3. Evaluate $\dfrac{\left(\lim\limits_{x\to 2} 3x^2+x+6\right)}{\left(\lim\limits_{x\to 1} 5x^2-2x+8\right)}$

4. If $f_x = x^2 + 2$ and $g_x = 7x + 4$, find $\dfrac{\lim\limits_{x \to 2} f_x}{\lim\limits_{x \to 2} g_x}$

5. Evaluate $\lim\limits_{x \to 1} \left(\dfrac{x^2 - 4x - 21}{x - 7} \right)$

6. Evaluate $\lim\limits_{x \to \alpha}\left(\dfrac{2x^3+2x^2+3x+2}{5x^3-2x^2+x-5}\right)$

7. Evaluate $\lim\limits_{x\to 3}(7x^4 - 3x^3 + 4x^2 - x - 1)$

8. Evaluate $\lim_{x \to 2}(2x^3 - x^2 + 5)$

9. Given that;

$$P_1 = 2x^2 + 5x + 6, \qquad P_2 = 3x^2 - 2x + 1$$

find $\lim_{x \to 0}(P_1)(P_2)$

10. Evaluate $\lim_{x \to 2}(x^2 + 2)(x + 3)$

11. Evaluate $\lim\limits_{x \to 1}(x^3 + 5)$

12. If $f_x = x^2 + 8x + 15$ and $g_x = x + 3$, find $\lim\limits_{x \to 2}\left(\dfrac{f_x}{g_x}\right)$

13. Evaluate $\lim\limits_{x \to \frac{1}{5}} \left(\dfrac{25x^2 - 4}{5x + 2} \right)$

14. Evaluate $\lim\limits_{x \to \alpha} \left(\dfrac{x^2 - 2x - 5}{2x^2 - 3x + 1} \right)$

15. Find the limiting value of $g_x = (x - 1)(2x + 1)$ as x approaches 2.

Chapter Three

L'HOSPITAL RULE

Limits solution using l'hospital's Rule

This rule can also be called l'hopital Rule. It is a rule that uses the derivatives to evaluate limits involving indeterminate forms. The limit is in form of $\left(\frac{f_x}{g_x}\right)$ and this either result into form $\frac{0}{0}, \frac{\infty}{\infty}$ or $\frac{n}{0}$. Whenever, we have this case, l'hospital rule is used to determine the limit of the functions.

Illustration

$$\lim_{x \to a}\left(\frac{f_x}{g_x}\right) = \lim_{x \to a}\left(\frac{f'_x}{g'_x}\right)$$

The numerator and denomination is differentiated separately.

In the function above

f'_x is the derivative of f_x

g'_x is the derivative of g_x

Note: the process of differentiation is applied until a definite result is obtained.

If $f_x = x^{an}$

To differentiate the function.

$$f'_x = ax^{n-1}$$

where **n** is an integer, the derivative of a constant is 0.

Example 1

Using L'hospital rule Evaluate:

$$\lim_{x \to 0} \left(\frac{2x^3 + 4x + 1}{x^2 + 5x} \right)$$

Solution

When you see a question in this form, and you are required to use l'hospital rule.

The first step to take before finding the limit of the function is to differentiate the numerator and the denominator.

That is; $\lim_{x \to a} \left(\frac{f_x}{g_x} \right) = \lim_{x \to a} \left(\frac{f'_x}{g'_x} \right)$

If $\lim_{x \to a} \left(\frac{f_x}{g_x} \right)$ is compared with $\lim_{x \to 0} \left(\frac{2x^3 + 4x + 1}{x^2 + 5x} \right)$

$$f_x = 2x^3 + 4x + 1$$

To differentiate the function above.

$$f_x = x^{an}$$

$$f'_x = ax^{n-1}$$

where n is an integer, the derivative of a constant is 0.

$$f_x = 2x^3 + 4x + 1$$

$$f'_x = 3(2x^{3-1}) + 4x^{1-1} + 0 = 6x^2 + 4$$

For g_x, $\qquad g_x = x^2 + 5x$

$$g'_x = 2 \times x^{2-1} + 5x^{1-1} = 2x + 5$$

$$\lim_{x \to a} \left(\frac{f_x}{g_x} \right) = \lim_{x \to a} \left(\frac{f'_x}{g'_x} \right)$$

$$\lim_{x \to 0} \left(\frac{2x^3 + 4x + 1}{x^2 + 5x} \right) = \lim_{x \to 0} \left(\frac{6x^2 + 4}{2x + 5} \right)$$

$$\lim_{x \to 0} \left(\frac{6x^2 + 4}{2x + 5} \right) = \frac{\lim_{x \to 0} (6x^2 + 4)}{\lim_{x \to 0} (2x + 5)}$$

Separate the functions using law 3

$$\frac{\lim_{x \to 0} 6x^2 + \lim_{x \to 0} 4}{\lim_{x \to 0} 2x + \lim_{x \to 0} 5}$$

Following law 2, separate the product of a constant and limit

of a function

$$\frac{6\lim_{x \to 0} x^2 + \lim_{x \to 0} 4}{2\lim_{x \to 0} x + \lim_{x \to 0} 5}$$

Replace x with 0, as x → 0

$$\frac{6(0) + 4}{2(0) + 5} = \frac{4}{5}$$

Therefore;

$$\lim_{x \to 0} \left(\frac{2x^3 + 4x + 1}{x^2 + 5x} \right) = \frac{4}{5}$$

Example 2

Using L'hospital rule Evaluate:

$$\lim_{x \to 1} \left(\frac{x^3 - 1}{x^2 - 1} \right)$$

$$\lim_{x\to a}\left(\frac{f_x}{g_x}\right) = \lim_{x\to a}\left(\frac{f'_x}{g'_x}\right)$$

If $\lim_{x\to a}\left(\frac{f_x}{g_x}\right)$ is compared with $\lim_{x\to 1}\left(\frac{x^3-1}{x^2-1}\right)$

Therefore;

$$f_x = x^3 - 1$$

$$f'_x = 3(x^{3-1}) - 0 = 3x^2 - 0 = 3x^2$$

For g_x ,

$$g_x = x^2 - 1$$

$$g'_x = 2 \times x^{2-1} - 0 = 2x - 0 = 2x$$

$$\lim_{x\to a}\left(\frac{f_x}{g_x}\right) = \lim_{x\to a}\left(\frac{f'_x}{g'_x}\right)$$

$$\lim_{x\to 1}\left(\frac{3x^2}{2x}\right) = \lim_{x\to 1}\left(\frac{3x^2}{2x}\right)$$

Separate the functions using law 3

$$\frac{\lim_{x\to 1} 3x^2}{\lim_{x\to 1} 2x} = \frac{\lim_{x\to 1} 3x}{\lim_{x\to 1} 2}$$

Following law 2, separate the product of a constant and limit of a function

$$\frac{3\lim_{x\to 1} x}{\lim_{x\to 1} 2}$$

Replace x with 0, as x → 1

$$\frac{3(1)}{2} = \frac{3}{2}$$

Therefore;

$$\lim_{x\to 1}\left(\frac{x^3-1}{x^2-1}\right) = \frac{3}{2}$$

Example 3

Evaluate $\displaystyle \lim_{x \to -3} \left(\frac{x^2 - 9}{(x+3)^2} \right)$

Solution

$$\lim_{x \to -3} \left(\frac{x^2 - 9}{(x + 3)^2} \right)$$

First of all, simplify the equation in the bracket

$$\lim_{x \to -3} \left(\frac{x^2 - 9}{(x + 3)^2} \right)$$

Recall,

$$x^2 - a^2 = (x - a)(x + a)$$

$$= \lim_{x \to -3} \left(\frac{(x - 3)(x + 3)}{(x + 3)(x + 3)} \right)$$

Note,

$$\frac{(x + 3)}{(x + 3)} = 1$$

$$= \lim_{x \to -3} \left(\frac{(x - 3)}{(x + 3)} \right)$$

Using l'hospital Rule

$$\lim_{x \to a} \left(\frac{f_x}{g_x} \right) = \lim_{x \to a} \left(\frac{f'_x}{g'_x} \right)$$

If $\displaystyle \lim_{x \to a} \left(\frac{f_x}{g_x} \right)$ *is compared with* $\displaystyle \lim_{x \to -3} \left(\frac{(x-3)}{(x+3)} \right)$

$$f_x = x - 3$$

$$f'_x = (x^{1-1}) - 0 = 1 - 0 = 1$$

For g_x,
$$g_x = x - 1$$
$$g'_x = x^{1-1} - 0 = 1 - 0 = 1$$
$$\lim_{x \to a} \left(\frac{f_x}{g_x}\right) = \lim_{x \to a} \left(\frac{f'_x}{g'_x}\right)$$
$$\lim_{x \to -3} \left(\frac{(x-3)}{(x+3)}\right) = \lim_{x \to -3} \left(\frac{1}{1}\right)$$

Separate the functions using law 3
$$\frac{\lim\limits_{x \to -3} 1}{\lim\limits_{x \to -3} 1} = \frac{\lim\limits_{x \to -3} 1}{\lim\limits_{x \to -3} 1} = 1$$

Therefore;
$$\lim_{x \to -3} \left(\frac{x^2-9}{(x+3)^2}\right) = 1$$

Example 4

Using L'hospital rule Evaluate:
$$\lim_{x \to 1} \left(\frac{\sqrt{x-1}}{x-1}\right)$$

Solution

$$\lim_{x \to a} \left(\frac{f_x}{g_x}\right) = \lim_{x \to a} \left(\frac{f'_x}{g'_x}\right)$$

If $\lim\limits_{x \to a} \left(\frac{f_x}{g_x}\right)$ is compared with $\lim\limits_{x \to 1} \left(\frac{\sqrt{x-1}}{x-1}\right)$

$$f_x = \sqrt{x} - 1 = (x)^{\frac{1}{2}} - 1$$

Differentiate f_x
$$f'_x = \frac{1}{2}x^{-\frac{1}{2}} - 0 = \frac{1}{2}x^{-\frac{1}{2}}$$
$$g_x = x - 1$$

Differentiate g_x

$$g'_x = 1 - 0 = 1$$

$$\lim_{x \to a} \left(\frac{f_x}{g_x} \right) = \lim_{x \to a} \left(\frac{f'_x}{g'_x} \right)$$

Separate the functions using law 3

$$\lim_{x \to 1} \left(\frac{\sqrt{x} - 1}{x - 1} \right) = \lim_{x \to 1} \left(\frac{\frac{1}{2} x^{-\frac{1}{2}}}{x} \right)$$

Following law 2, separate the product of a constant and limit

of a function

$$= \frac{\frac{1}{2} \lim_{x \to 1} x^{-\frac{1}{2}}}{\lim_{x \to 1} x}$$

Replace x with 0, as x → 1

$$\frac{\frac{1}{2}(1)}{1} = \frac{1}{2}$$

Therefore;

$$\lim_{x \to 1} \left(\frac{\sqrt{x} - 1}{x - 1} \right) = \frac{1}{2}$$

Example 5

Using L'hospital rule Evaluate:

$$\lim_{x \to 0} \left(\frac{\sqrt[3]{x+27}}{x} \right)$$

Solution

$$\lim_{x \to a} \left(\frac{f_x}{g_x} \right) = \lim_{x \to a} \left(\frac{f'_x}{g'_x} \right)$$

If $\lim_{x \to a} \left(\frac{f_x}{g_x} \right)$ is compared with $\lim_{x \to 0} \left(\frac{\sqrt[3]{x+27}}{x} \right)$

Then,

$$f_x = \sqrt[3]{x + 27} = (x + 27)^{\frac{1}{3}}$$

Differentiate f_x

$$f'_x = \frac{1}{3}(1)(x + 27)^{\frac{1}{3} - 1} = \frac{1}{3}(x + 27)^{-\frac{2}{3}}$$

$$g_x = x$$

Differentiate g_x

$$g'_x = 1$$

$$\lim_{x \to a} \left(\frac{f_x}{g_x} \right) = \lim_{x \to a} \left(\frac{f'_x}{g'_x} \right)$$

$$\lim_{x \to 0} \left(\frac{\sqrt[3]{x + 27}}{x} \right) = \lim_{x \to 0} \left(\frac{\frac{1}{3}(x + 27)^{-\frac{2}{3}}}{1} \right)$$

Separate the functions using law 3

$$\lim_{x \to 0} \frac{1}{3}(x + 27)^{-\frac{2}{3}}$$

Following law 2, separate the product of a constant and limit of a function

$$\frac{1}{3} \lim_{x \to 0}(x + 27)^{-\frac{2}{3}}$$

Replace x with 0, as x $\to$ 0

$$\frac{1}{3}(0+27)^{\frac{2}{3}} = \frac{1}{3}(27)^{-\frac{2}{3}}$$

$$\frac{1}{3}(3^3)^{-\frac{2}{3}} = \frac{1}{3}(3)^{-2} = \frac{1}{3} \times \frac{1}{9} = \frac{1}{27}$$

Therefore;

$$\lim_{x \to 0}\left(\frac{\sqrt[3]{x+27}}{x}\right) = \frac{1}{27}$$

Exercise 3: Try this on your own

Using l'hospital rule,

1. Evaluate $\lim_{x \to a}\left(\frac{x^3-a^3}{x-a}\right)$

2. Evaluate $\lim\limits_{x \to a} \left(\dfrac{x^2 + 4ax - 5a^2}{x - a} \right)$

3. Evaluate $\lim\limits_{x \to 7} \left(\dfrac{2 - \sqrt{x-3}}{x^2 - 49} \right)$

4. *Evaluate* $\lim\limits_{x \to 4} \left(\dfrac{x^3 - 64}{x - 4} \right)$

Chapter Four

CONTINUITY OF A FUNCTION

A function is continuous at $x = a$ if the following conditions are met;

a) The function is defined at $x = a$. This means
   $f_{(a)}$ *equals a real number.* $f_{(a)}$ is defined.

b) The limit of the function exists as x approaches a. That means $\lim\limits_{x \to a} f_x$ exist.

c) The limit of the function as x approaches a is equal to the function value at $x = a$.

   that is; $\lim\limits_{x \to a} f_x = f_{(a)}$.

with conditions stated above, you can know maybe a function is continuous or discontinuous.

Example 1

Determine whether f_x is continuous at $x = 2$ and at $x = -1$.

$$f_x = \frac{x^2 - x - 2}{x + 1}$$

Solution

i.   How do we know f_x is continuous at $x = 2$?

First condition

$$f_x \text{ must be defined at at } x = 2$$

$$f_2 = \frac{2^2 - 2 - 2}{2 + 1} = \frac{0}{3} = 0$$

$$f_2 = 0$$

Since $f_2 = 0$, the function f_2 **is** defined. The first condition is met.

Second condition

The limit of the function f_x must exist as x approaches 2.

$$\lim_{x \to a} f_x = \lim_{x \to 2} \frac{x^2 - x - 2}{x + 1}$$

$$\frac{2^2 - 2 - 2}{2 + 1} = 0$$

The limit, $\lim_{x \to 2} \frac{x^2 - x - 2}{x + 1}$ exist at 0. This made the second condition valid for this function.

Third condition

The limit of function as x approaches 2 must be equal to the function value at $x = 2$.

That is;

$$\lim_{x \to 2} f_x = f_2$$

From the analysis $\lim_{x \to 2} f_x = 0$ and $f_2 = 0$. This makes the condition valid.

Since the three conditions are met for the function

$$f_x = \frac{x^2 - x - 2}{x + 1}$$

as x approaches **2**.

We can conclude that f_x is continuous at $x = 2$.

ii.   How do we know f_x is continuous at $x = -1$.

First condition

$$f_x \text{ must be defined at at } x = -1$$

$$f_{-1} = \frac{-1^2 - (-1) - 2}{-1 + 1} = \frac{0}{0}$$

$$f_{-1} = \frac{0}{0}$$

Since $f_{-1} = \frac{0}{0}$, the function f_{-1} **is not** defined. The first condition is not met.

Since the first condition is not met. The function f_x is discontinuous at $x = -1$.

Example 2

Determine whether f_x is continuous at at $x = 1$.

$$f_x = \frac{x^3 - 1}{x - 1}$$

Solution

i.   How do we know f_x is continuous at $x = 1$?

First condition

$$f_x \text{ must be defined at at } x = 1$$

$$f_1 = \frac{1^3 - 1}{1 - 1} = \frac{0}{0}$$

$$f_{-1} = \frac{0}{0}$$

Since $f_1 = \frac{0}{0}$, the function f_1 **is not** defined. The first condition is not met.

Since the first condition is not met. The function f_x is discontinuous at $x = 1$.

Example 3

Determine whether f_x is continuous at at $x = 2$ and at $x = -1$.

$$f_x = \begin{vmatrix} \dfrac{x^3 - 1}{x - 1} & x \neq 1 \\ 2 & x = 1 \end{vmatrix}$$

Solution

i.   How do we know f_x is continuous at $x = 1$?

First condition

$$f_x \text{ must be defined at at } x = 1$$

$$f_1 = 2$$

Since $f_1 = 2$, the function f_1 **is** defined. The first condition is met.

Second condition

The limit of the function f_x must exist as x approaches 1.

$$\lim_{x \to a} f \quad x = \lim_{x \to 1} \frac{x^3 - 1}{x - 1}$$

Note

$$a^3 - b^3 = (a - b)(a^2 + ab + b^2)$$

Therefore:

$$x^3 - 1 = x^3 - 1^3$$

$$x^3 - 1^3 = (x - 1)(x^2 + x + 1)$$

$$\lim_{x \to 1} \frac{x^3 - 1}{x - 1} = \lim_{x \to 1} \frac{(x - 1)(x^2 + x + 1)}{x - 1}$$

$$\lim_{x \to 1} x^2 + x + 1 = 1 + 1 + 1 = 3$$

$$\lim_{x \to 1} \frac{x^3 - 1}{x - 1} = 3$$

The limit $\lim_{x \to 1} \frac{x^3-1}{x-1}$ exist at 3. This made the second condition

valid for this function.

Third condition

The limit of function as x approaches 1 *must be equal to the*

function value at $x = 1$.

That is;

$$\lim_{x \to 1} f_x = f_1$$

From the analysis $\lim_{x \to 1} f_x = 3$ and $f_1 = 2$. This condition is not

valid as;

$$\lim_{x \to 1} f_x \neq f_1$$

Since the third condition is not met.

The function f_x is discontinuous at $x = 1$.

Exercise 4: Try this on your own

1. Examine $f_x = 7x^2 + 3x + 8$ for continuity at $x = -1$.

2. Examine $h_x = 4x^3 + 3x^2 + 2x - 1$ for continuity at $x = 2$.

Chapter Five

LIMIT OF TRIGONOMETRY FUNCTION

In this chapter, we shall look on how limit of a trigonometric function can be found.

To evaluate the limit of a trigonometry function, take note of the following functions.

- $\lim\limits_{x \to 0} \left(\dfrac{\sin x}{x} \right) = 1$

- $\lim\limits_{x \to 0} \left(\dfrac{\tan x}{x} \right) = 1$

Where x is an angle in radian.

For continuity, sine and cosine function are continuous everywhere, such that;

- $\lim\limits_{x \to a} \sin x = \sin a$

- $\lim\limits_{x \to a} \cos x = \cos a$

- $\lim\limits_{x \to a} \tan x = \tan x$

- $\lim\limits_{x \to a} \cot x = \cot a$

- $\lim\limits_{x \to a} \sec x = \sec a$

- $\lim\limits_{x \to a} \csc x = \csc a$

Example 1

Evaluate

$$\lim_{x \to \pi} \frac{\tan x}{\sin 2x}$$

Solution

From trigonometry;

$$\tan x = \frac{\sin x}{\cos x}$$

$$\sin 2x = \sin(x + x)$$

$$\sin(x + x) = \sin x \cos x + \cos x \sin x$$

$$= 2 \sin x \cos x$$

$$\frac{\tan x}{\sin x} = \frac{\frac{\sin x}{\cos x}}{2 \sin x \cos x}$$

$$= \frac{\sin x}{\cos x \,(2\sin x \cos x)}$$

$$= \frac{\sin x}{2 \sin x \cos^2 x}$$

$$= \frac{1}{2 \cos^2 x}$$

Therefore;

$$\lim_{x \to \pi} \frac{\tan x}{\sin 2x} = \lim_{x \to \pi} \frac{1}{2 \cos^2 x}$$

$$\lim_{x \to \pi} \frac{1}{2 \cos^2 x} = \frac{1}{2} \lim_{x \to \pi} \frac{1}{\cos^2 x}$$

$$\frac{1}{2}\left(\frac{1}{\cos^2 \pi}\right)$$

Where ;

$$\pi = 180$$

$$\frac{1}{2}\left(\frac{1}{\cos^2 180}\right) = \frac{1}{2}\left(\frac{1}{(-1)^2}\right) = \frac{1}{2}$$

Example 2

Evaluate

$$\lim_{x \to 0} \frac{12x - 4\sin x}{x}$$

Solution

$$\lim_{x \to 0} \frac{12x - 4\sin x}{x}$$

Separate the fraction

$$\lim_{x \to 0} \left(\frac{12x}{x} - \frac{4\sin x}{x}\right)$$

If $\lim_{x \to a}(f_x \pm g_x) = \lim_{x \to a} f_x \pm \lim_{x \to a} g_x$

Then;

$$\lim_{x \to 0} \left(\frac{12x}{x} - \frac{4\sin x}{x}\right) = \lim_{x \to 0} \frac{12x}{x} - \lim_{x \to 0} \frac{4\sin x}{x}$$

$$\lim_{x \to 0} \frac{12x}{x} - 4\lim_{x \to 0} \frac{\sin x}{x}$$

$$\lim_{x \to 0} 12 - 4\lim_{x \to 0} \frac{\sin x}{x}$$

Recall;

$$\lim_{x \to 0} \left(\frac{\sin x}{x} \right) = 1$$

$$\lim_{x \to 0} 12 - 4 \lim_{x \to 0} \frac{\sin x}{x} = 12 - 4(1) = 8$$

Therefore;

$$\lim_{x \to 0} \frac{12x - 4 \sin x}{x} = 8$$

Example 3

Evaluate

$$\lim_{x \to 0} \frac{\tan x}{x}$$

Solution

$$\lim_{x \to 0} \frac{\tan x}{x}$$

If $\tan x = \frac{\sin x}{\cos x}$

$$\lim_{x \to 0} \frac{\tan x}{x} = \lim_{x \to 0} \frac{\sin x / \cos x}{x}$$

$$\lim_{x \to 0} \frac{\sin x}{x \cos x} = \lim_{x \to 0} \frac{1}{\cos x} \cdot \frac{\sin x}{x}$$

$$= \left(\lim_{x \to 0} \frac{1}{\cos x} \right) \left(\lim_{x \to 0} \frac{\sin x}{x} \right)$$

Recall;

$$\lim_{x \to 0} \left(\frac{\sin x}{x} \right) = 1$$

Then;

$$\left(\frac{1}{\cos 0}\right)(1) = \left(\frac{1}{1}\right)(1) = 1$$

$$\lim_{x \to 0} \frac{\tan x}{x} = 1$$

Example 4

Evaluate

$$\lim_{x \to 0} \frac{\cos x}{x}$$

Solution

$$\lim_{x \to 0} \frac{\cos x}{x}$$

If $\tan x = \frac{\sin x}{\cos x}$, therefore $\cos x = \frac{\sin x}{\tan x}$

$$\lim_{x \to 0} \frac{\cos x}{x} = \lim_{x \to 0} \frac{\sin x / \tan x}{x}$$

$$\lim_{x \to 0} \frac{\sin x}{x \tan x} = \lim_{x \to 0} \frac{1}{\tan x} \cdot \frac{\sin x}{x}$$

$$= \left(\lim_{x \to 0} \frac{1}{\tan x}\right)\left(\lim_{x \to 0} \frac{\sin x}{x}\right)$$

Recall;

$$\lim_{x \to 0} \left(\frac{\sin x}{x} \right) = 1$$

Then;

$$\left(\frac{1}{\tan 0} \right)(1) = \left(\frac{1}{0} \right)(1) = no\ Limit.$$

$$\lim_{x \to 0} \frac{\cos x}{x} = no\ limit$$

Example 5

Evaluate

$$\lim_{x \to 0} \frac{\sin 5x}{x}$$

Solution

$$\lim_{x \to 0} \frac{\sin 5x}{x}$$

Rewrite the function;

$$\frac{\sin 5x}{x} = \frac{5 \sin 5x}{5x}$$

Therefore;

$$\lim_{x \to 0} \frac{\sin 5x}{x} = \lim_{x \to 0} \frac{5 \sin 5x}{5x}$$

If $y = 5x$ when $x \to 0, y = 0$

Substitute $5x$ for y;

$$\lim_{x \to 0} \frac{5 \sin 5x}{5x} = \left(\lim_{x \to 0} 5 \right) \left(\lim_{y \to 0} \frac{\sin y}{y} \right)$$

Note:

As $\lim\limits_{x\to 0}\dfrac{\sin x}{x} = 1,\ $ so $\lim\limits_{y\to 0}\dfrac{\sin y}{y} = 1$

$$\left(\lim\limits_{x\to 0} 5\right)\left(\lim\limits_{y\to 0}\dfrac{\sin y}{y}\right) = 5(1) = 5$$

$$\lim\limits_{x\to 0}\dfrac{\sin 5x}{x} = 5$$

Example 6

Evaluate

$$\lim\limits_{t\to 0}\dfrac{\sin 3t}{4t}$$

Solution

$$\lim\limits_{t\to 0}\dfrac{\sin 3t}{4t}$$

Rewrite the function;

$$\dfrac{\sin 3t}{4t} = \dfrac{3\sin 3t}{4(3t)}$$

Therefore;

$$\lim\limits_{t\to 0}\dfrac{\sin 3t}{4t} = \lim\limits_{t\to 0}\dfrac{3}{4}\cdot\dfrac{\sin 3t}{3t}$$

If $y = 3t$ when $t \to 0, y = 0$

Substitute $3t$ for y;

$$\lim\limits_{t\to 0}\dfrac{3}{4}\cdot\dfrac{\sin 3t}{3t} = \left(\lim\limits_{t\to 0}\dfrac{3}{4}\right)\left(\lim\limits_{y\to 0}\dfrac{\sin y}{y}\right)$$

Note:

As $\lim\limits_{t\to 0}\dfrac{\sin t}{t} = 1$, so $\lim\limits_{y\to 0}\dfrac{\sin y}{y} = 1$

$$\left(\lim_{t\to 0}\frac{3}{4}\right)\left(\lim_{y\to 0}\frac{\sin y}{y}\right) = \frac{3}{4}(1) = \frac{3}{4}$$

$$\lim_{t\to 0}\frac{\sin 3t}{4t} = \frac{3}{4}$$

Example 7

Evaluate

$$\lim_{x\to 1}\frac{\sin(x-1)}{x^2+2x-3}$$

Solution

$$\lim_{x\to 1}\frac{\sin(x-1)}{x^2+2x-3}$$

Factorize the denominator

$$x^2 + 2x - 3 = x^2 + 3x - x - 3$$
$$x(x+3) - 1(x+3)$$
$$(x+3)(x-1)$$

Therefore;

$$\lim_{x\to 1}\frac{\sin(x-1)}{x^2+2x-3} = \lim_{x\to 1}\frac{\sin(x-1)}{(x+3)(x-1)}$$

$$=\lim_{x\to 1}\frac{1}{x+3}\cdot\frac{\sin(x-1)}{x-1}$$

If $y = x - 1$ when $x \to 1, y = 0$

Substitute $x - 1$ for y;

$$\lim_{x \to 1} \frac{1}{x + 3} \cdot \frac{\sin(x - 1)}{x - 1} = \left(\lim_{x \to 1} \frac{1}{x + 3}\right)\left(\lim_{y \to 0} \frac{\sin y}{y}\right)$$

Note:

As $\lim_{x \to 0} \frac{\sin x}{x} = 1$, so $\lim_{y \to 0} \frac{\sin y}{y} = 1$

$$\left(\lim_{x \to 1} \frac{1}{x + 3}\right)\left(\lim_{y \to 0} \frac{\sin y}{y}\right) = \frac{1}{1 + 3}(1) = \frac{1}{4}$$

$$\lim_{x \to 1} \frac{\sin(x - 1)}{x^2 + 2x - 3} = \frac{1}{4}$$

Example 8

Evaluate

$$\lim_{x \to a} \left(x \sin \frac{2}{\pi}\right)$$

Solution

$$\lim_{x \to a} \left(x \sin \frac{2}{\pi}\right)$$

Rewrite the function;

$$x \sin \frac{2}{\pi} = \frac{2 \sin \frac{2}{\pi}}{\frac{2}{\pi}}$$

Therefore;

$$\lim_{x \to a} \left(x \sin \frac{2}{\pi}\right) = \lim_{x \to a} \frac{2 \sin \frac{2}{\pi}}{\frac{2}{\pi}}$$

$$= \left(\lim_{x \to \alpha} 2\right)\left(\lim_{x \to \alpha} \frac{\sin\frac{2}{\pi}}{\frac{2}{\pi}}\right)$$

If $y = \frac{2}{\pi}$ when $x \to \pi, y = 0$

Substitute $\frac{2}{\pi}$ for y;

$$\left(\lim_{x \to \alpha} 2\right)\left(\lim_{x \to \alpha} \frac{\sin\frac{2}{\pi}}{\frac{2}{\pi}}\right) = \left(\lim_{x \to \alpha} 2\right)\left(\lim_{y \to 0} \frac{\sin y}{y}\right)$$

Note:

As $\lim_{x \to 0} \frac{\sin x}{x} = 1$, so $\lim_{y \to 0} \frac{\sin y}{y} = 1$

$$\left(\lim_{x \to \alpha} 2\right)\left(\lim_{y \to 0} \frac{\sin y}{y}\right) = 2(1) = 2$$

$$\lim_{x \to \alpha}\left(x \sin\frac{2}{\pi}\right) = 2$$

Example 9

Evaluate

$$\lim_{t \to \frac{\pi}{4}}(t \tan t)$$

Solution

$$\lim_{t \to \frac{\pi}{4}}(t \tan t) = \frac{\pi}{4}\tan\frac{\pi}{4}$$

Note; in degree $\pi = 180^0$

Therefore;

$$= \frac{\pi}{4}\tan\frac{180}{4}$$

$$= \frac{\pi}{4}\tan 45$$

$$= \frac{\pi}{4}(1)$$

$$\frac{\pi}{4}$$

$$\lim_{t\to\frac{\pi}{4}}(t\tan t) = \frac{\pi}{4}$$

Example 10

Evaluate

$$\lim_{x\to\frac{\pi}{2}} \frac{\cos^2 x}{1-\sin x}$$

Solution

If we substitute $\frac{\pi}{2}$ directly into the function, the denominator will be zero (0)which will make the limit not exist

From trigonometry function;

$$\sin^2 x + \cos^2 x = 1$$
$$\cos^2 x = 1 - \sin^2 x$$

Substitute $1 - \sin^2 x$ for $\cos^2 x$

$$\lim_{x\to\frac{\pi}{2}} \frac{\cos^2 x}{1-\sin x} = \lim_{x\to\frac{\pi}{2}} \frac{1-\sin^2 x}{1-\sin x}$$

If $a^2 - b^2 = (a-b)(a+b)$

$$\lim_{x\to\frac{\pi}{2}} \frac{1-\sin^2 x}{1-\sin x} = \lim_{x\to\frac{\pi}{2}} \frac{1^2-\sin^2 x}{1-\sin x}$$

$$1^2 - \sin^2 x = (1-\sin x)(1+\sin x)$$

$$\lim_{x\to\frac{\pi}{2}}\frac{1^2-\sin^2 x}{1-\sin x} = \lim_{x\to\frac{\pi}{2}}\frac{(1-\sin x)(1+\sin x)}{1-\sin x}$$

$$\lim_{x\to\frac{\pi}{2}}\frac{(1-\sin x)(1+\sin x)}{1-\sin x} = \lim_{x\to\frac{\pi}{2}}(1+\sin x)$$

$$\lim_{x\to\frac{\pi}{2}}(1+\sin x) = \lim_{x\to\frac{\pi}{2}}(1) + \lim_{x\to\frac{\pi}{2}}(\sin x)$$

$$\lim_{x\to\frac{\pi}{2}}(1) + \lim_{x\to\frac{\pi}{2}}(\sin x) = 1 + \sin\frac{\pi}{2}$$

Note; in degree $\pi = 180^0$

$$= 1 + \sin\frac{180}{2} = 1 + \sin 90$$

$$= 1 + \sin 90 = 1 + 1 = 2$$

$$\lim_{x\to\frac{\pi}{2}}\frac{\cos^2 x}{1-\sin x} = 2$$

EXERCISE 5

Find the limit of the following functions

1. *Evaluate* $\lim_{x\to\frac{\pi}{2}}\dfrac{\sin x}{\cos 6x}$

2. *Evaluate* $\lim\limits_{x \to 0} \dfrac{\sin 3x}{x}$

3. Evaluate $\lim\limits_{t \to 0} \dfrac{\sin 3\, t}{\cos 2t}$

4. Evaluate $\lim\limits_{t \to 2} \dfrac{\sin(t-2)}{t^2-t-2}$

5. *Evaluate* $\displaystyle\lim_{x\to 0} x \cot 2x$

6. *Evaluate* $\lim\limits_{t \to 2} \dfrac{\sin(x+3)}{x^3+3x^2}$

7. Evaluate $\displaystyle\lim_{t\to 0}\frac{\cos 3t}{\cos 2t}$

A N S W E R S

Exercise 2

1. $-3 + a$

2. 4

3. $\frac{20}{11}$

4. $\frac{1}{3}$

5. 4

6. $\frac{2}{5}$

7. 518

8. 17

9. 6

10. 30

11. 6

12. 7

13. -1

14. $\frac{1}{2}$

15. 5

Exercise 3

1. $3a^2$

2. $6a$

3. $\frac{1}{2}$

4. 5

Exercise 4.

1. f_x is continous
2. h_x is continous

Exercise 5

1. -1
2. 3
3. 0
4. $\frac{1}{3}$
5. $\frac{1}{2}$
6. $\frac{1}{9}$
7. 1

Improve your Math Skills with other Books

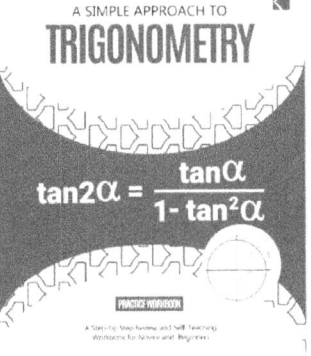

www.ingramcontent.com/pod-product-compliance
Lightning Source LLC
Chambersburg PA
CBHW060349130626
46553CB00003B/1147